THE SINAGUA

by Christian E. Downum

WHO WERE THE SINAGUA?	2
HOW DID THE SINAGUA LIVE?	4
HISTORY OF THE SINAGUA	13
CONCLUSION	31

MUSEUM OF NORTHERN ARIZONA

WHO WERE THE SINAGUA?

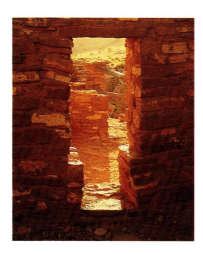

Keyhole doorway, Lomaki ruin, Wupatki National Monument. Photograph by Ralph Lee Hopkins

FROM the eighth through the fifteenth centuries A.D., the canyons, grasslands, mountains, and mesas of central and northern Arizona were home to a creative and resilient people called the Sinagua. At the height of their influence and geographical extent during the twelfth and thirteenth centuries, they lived throughout most of the San Francisco Mountains volcanic field and much of the upper and middle Verde River Valley. Their culture was a synthesis of surrounding cultural elements—borrowed, adapted, and reconfigured into an eclectic blend. For this reason, the Sinagua developed a lifestyle that was similar to that of their neighbors but one that was always recognizably distinct from them.

The origins of the Sinagua are something of a mystery. We know that people have occupied central and northern Arizona for at least 11,500 years because their artifacts are relatively abundant. Much of this evidence, however, is in the form of isolated projectile points or poorly dated artifacts scattered on the surface. Moreover, the exact relationship between these early people and subsequent ceramic-using cultures is unknown. For these reasons, we simply do not know whether the Sinagua culture was created by people already present or whether the first Sinagua immigrated from some other area.

Early Sinagua sites share a number of similarities with those of the Mogollon culture to the south and east. Still, we do not know if the Mogollon people themselves immigrated or if their traits were adopted by people already living in Sinagua territory.

The answer to the question "Who were the Sinagua?" has three dimensions: time, geography, and material culture. The first two we can treat rather straightforwardly. The Sinagua emerged as a distinct culture about A.D. 700 and disappeared sometime around A.D. 1450, when they abandoned their heartland and with others retreated to the Hopi mesas. Geographically, the Sinagua occupied an area bounded roughly by the Little Colorado River, the San Francisco Peaks, the Verde River, and the East Verde River. Within this territory, archaeologists have traditionally identified at least two divisions: the northern Sinagua, living above the Mogollon Rim, and the southern Sinagua, inhabiting the middle and upper reaches of the Verde River Valley and the territory drained by its tributaries.

The third category is more problematical. The objects used by the archaeologist to define a culture (such as pottery, architecture, stone tools, clothing, and ceremonial items) are referred to as "material culture." Since Sinagua material culture varied greatly, it is difficult to isolate items common to them. Perhaps the only truly definitive element was Alameda Brown Ware pottery, a nondecorated ware made from volcanic clays and manufactured with a technique called paddle-and-anvil.

Other than these rather loose criteria, defining the Sinagua is a difficult proposition. Throughout their history and across their vast territory, Sinagua subsistence, settlement, organization, architecture, artifacts, and burial practices varied greatly. There were, evidently, many ways to be Sinagua.

Wukoki ruin, Wupatki National Monument. Photograph by Tom Danielsen

HOW DID THE SINAGUA LIVE?

MOST of us have basic questions about people separated from us by time: How did these people make a living? What did they eat? What kinds of houses did they build, and what kinds of objects did they use in their everyday and ceremonial life? Much of what we know about the Sinagua, unfortunately, is incomplete and frustratingly uneven. For some categories of behavior, time periods, and regions, we have many details. For others, however, there are glaring gaps in our knowledge, and we are forced to admit that the Sinagua still hold important secrets. Nonetheless, we can present a general picture of Sinagua life with some confidence.

Making a Living

Although specific patterns were highly variable, the Sinagua always made their living from some combination of hunting, gathering, and farming. Major game animals were pronghorn antelope, jackrabbits, cottontails, mule deer, and bighorn sheep. Woodrats, deer mice, prairie dogs, squirrels, and many other small species rounded out the list of meat sources. Large animals probably were taken with a throwing spear or bow and arrow. Hunters almost certainly used a variety of techniques and tools on smaller animals—including arrows, throwing sticks, clubs, and snares. Jackrabbits may have been bagged in communal "drives," in which the animals were chased into an arroyo or other natural trap strung with a long fiber net.

Wild plant foods were also a significant part of the Sinagua diet. A minimum list of staples would have included piñon nuts, grass seeds, prickly pear and cholla fruit, greens and seeds from annual plants such as goosefoot and amaranth, agave hearts, and the berries and leaves of many shrubs. The distribution of wild edibles varied with the season and the specific environment. Since different plants would have been available at various times of the year, the Sinagua probably scheduled their plant-gathering rounds accordingly.

Perhaps the most important aspect of Sinagua subsistence was agriculture. Although the dry climate and short growing season make many areas of Sinagua country extremely risky for cultivation, the Sinagua were extraordinarily clever and successful farmers. For more than 700 years, they managed to meet the challenges of nearly every environmental zone of central and northern Arizona with a myriad of small-scale but highly effective farming techniques, such as irrigation, floodwater farming at the mouths of arroyos, and dry farming.

Thus, no Sinagua population ever seems to have relied exclusively on hunting and gathering or on agriculture alone. Instead, they made diversity their strength. Much of this diversity was simply a response to the environment, which dictated the availability of natural foods and the reliability of agriculture in any particular place. Much, however, was due also to a conscious desire by the Sinagua to maintain a number of options as a hedge against the failure of any one food source. Such a strategy was essential in their marginal environment for it provided the foundation upon which rested all remaining aspects of Sinagua life.

Despite our general understanding of the importance of agriculture to the Sinagua, we still have much to learn about their specific farming practices and crops. Three recent studies of Sinagua fields have helped to show the wide range of strategies and techniques they cleverly tailored to specific environments.

One study was conducted by National Park Service Archaeologist Scott Travis at Wupatki National Monument. This area was a prehistoric frontier zone where Sinagua, Kayenta Anasazi, and Cohonina people settled and farmed during the twelfth and thirteenth centuries A.D. Using data from a recent monument-wide archaeological survey, Travis found that agriculture was practiced on a massive scale at Wupatki. In fact, the survey team recorded more than 12,000 individual features including rock piles, alignments, checkdams, and terraces. Not surprisingly, nearly three-fourths of the features were concentrated in the upland areas, which are wetter and have better soils.

A short distance south of Wupatki, researchers from Northern Arizona University studied an entirely different set of fields. High-altitude aerial thermographs (special photographs showing differences in the temperature of the ground surface) revealed a set of ridged fields created by piling volcanic cinders into long ridges and mounds. These constructions helped retain moisture in the underlying soil. By capturing solar radiation during the day, the rocks may have slowed nighttime cooling of the ground surface, effectively extending the growing season. Soil samples from some of these fields produced pollen from corn and a number of "disturbance" plants usually associated with cultivation. Other fields, however, failed to produce any pollen from domesticates, indicating they probably had been used only for cultivating wild plants such as goosefoot and amaranth.

Finally, in yet a different setting, Paul and Suzanne Fish, then of the Museum of Northern Arizona, studied Sinagua fields along Beaver Creek, a tributary of the Verde River. These fields, watered with irrigation ditches leading from the creek, were marked by a complex grid of stone alignments. Many small, single-room, masonry structures called field houses bordered the fields. Soil samples contained the pollen of corn and cotton. The Fishes calculated this system could have provided about 26 metric tons of corn in a year—enough food for about 127 people.

These three examples illustrate the extremely diverse farming strategies and techniques that the Sinagua employed. Obviously, practices varied widely depending on local topography and soil, moisture, and temperature conditions. As is so

Gathering agave.
Drawing by Denny Carley

Some of the wild food plants gathered by the Sinagua. Photograph by Denny Carley

characteristic of the Sinagua, a mosaic of patterns was cleverly tailored to specific environmental settings.

GETTING A DRINK

Where did the Sinagua get their drinking water? Even their name—from an early Spanish characterization of the San Francisco Mountains as the *Sierra Sin Agua* or "Waterless Mountains"— poses this puzzle. The sparse, uneven distribution of permanent water contrasts sharply with the widespread prehistoric occupation of this area. Literally thousands of pueblos and pithouses were located seemingly without regard for this daily necessity of human life.

In some cases, the solution to getting a drink is obvious. Permanent surface water can be found in Sinagua country, and many of the largest settlements were clustered around these sources. The southern Sinagua pueblo of Tuzigoot, for example, was built near the banks of the Verde River, and Montezuma Castle overlooks Beaver Creek, a permanent stream that empties into the Verde. Wupatki, Kinnikinnick, Chavez Pass, and many other northern Sinagua settlements also were built near reliable springs.

However, the majority of Sinagua sites are relatively far from the nearest stream, spring, or seep. Some lie many miles from the nearest plausible source of permanent water. How did these communities manage to exist? There is no single answer.

Nearly all settlements made use of seasonal rainfall, collecting it in bedrock depressions or "potholes," stream bottoms, and other natural cisterns. Ceramic vessels sometimes were used to capture rainwater; one example found near Wupatki Pueblo had been strategically placed below the rim of a sandstone outcrop to collect water flowing off the rock during heavy downpours. To supplement such natural sources, the Sinagua also constructed shallow reservoirs. Best termed "wet weather ponds," these reservoirs were formed by damming small drainages with

The people of Tuzigoot, a southern Sinagua pueblo, used the nearby Verde River as a permanent water source. Photograph by Dick Dietrich

Left: A century plant (agave) on the Mogollon Rim. Photograph by Ralph Lee Hopkins
Below: Farming was less precarious in warmer areas like Palatki. Photograph by Gražina Sakalas

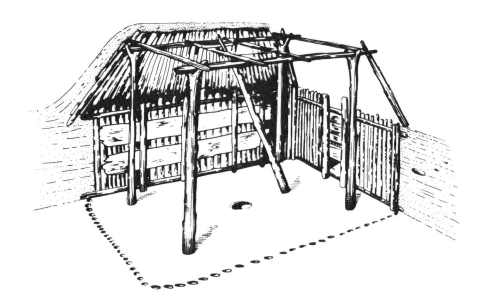

The Sinagua usually built a deep pithouse lined with wood or stone. Drawing by Pamela Lungé, after Colton

boulders and earth. They probably would have held water throughout the summer and winter rainy seasons and perhaps for a few weeks beyond.

What the Sinagua did when these ponds were dry is unknown, but two possibilities seem most likely. First, they could have moved to settlements closer to permanent water, shifting with the seasons or climate. Such mobility is relatively common in the ethnographic record of the native Southwest. Second, the Sinagua may have regularly made long-distance treks to the nearest reliable water source. This pattern also is well documented. In fact, some farming groups have been known to walk up to nine miles on a regular basis during the dry season to obtain drinking water. Whichever of these two options was pursued, the Sinagua were clearly capable of obtaining an adequate supply of drinking water.

Architecture

One of the most impressive aspects of the Sinagua culture is the tremendous variety of domestic architectural forms. The longevity of the culture probably accounts for many of these differences, but we also find considerable diversity in structures built during the same period. Although we do not always know the reasons for such diversity, it seems likely that particular environmental conditions, seasonal occupation, specialized functions, local availability of building materials, and family or local cultural traditions all played a role.

Subterranean structures were always part of the Sinagua architectural inventory. Before Sunset Crater erupted in A.D. 1064, most dwellings were built with floors beneath the ground surface. These varied in depth; some, considered to be true "pithouses," were built entirely below the ground, while others were merely "houses-in-a-pit"—that is, largely above-ground constructions placed in a shallow pit. Early on, Sinagua pithouses were made mostly of earth and wood, often with upright poles or split pine planks lining the insides of the house pit. Later, particularly after Sunset Crater's eruption, pithouses were lined with masonry.

Throughout Sinagua times, entry into pithouses was highly variable. Some buildings were entered through ramps or vestibules along the side, but many had holes in the roofs with a ladder or notched pole leading down. Nearly all of them had a hearth, usually located near the center of the room.

Through much of their history, the Sinagua constructed multiroomed, above-ground masonry pueblos. Building materials included sandstone and limestone slabs, basalt blocks, and cobbles. Mud mortar provided the cement, and the walls were topped with a roof supported by juniper, pine, fir, or cottonwood beams, sometimes brought in from several miles away. The majority of these structures consisted of only a few rooms. After about A.D. 1120, however, the Sinagua constructed some larger pueblos. Patterns of wall abutment, tree-ring dates, and other evidence indicate that these pueblos grew in increments, with only a few

Above: Wukoki ruin at Wupatki National Monument is an example of the multiroomed, above-ground masonry pueblos built by both the Sinagua and the Anasazi. Photograph by Gene Balzer

Right: Prairie dogs were a supplemental food source for the highly adaptable and resourceful Sinagua. Photograph by Damon Bullock

rooms added at any one time. Very late in the Sinagua sequence, some extraordinarily large pueblos were built—particularly in the Verde Valley and on Anderson Mesa. These were truly impressive structures of up to several hundred individual rooms stacked up to three stories high.

A final category of Sinagua domestic architecture was the cliff dwelling. These structures, dating mostly after A.D. 1100, are scattered widely throughout Sinagua country. Two national monuments, Montezuma Castle and Walnut Canyon, preserve some of the region's most picturesque examples. However, many other Sinagua cliff dwellings can be found throughout the area, particularly in the canyons of the Verde Valley.

Early archaeologists once thought that the "cliff dwellers" might have been a group distinct from the people who lived in open-air pueblos. All available evidence, however, shows this was not the case. It does seem possible that cliff dwellings served special functions. Perhaps they were occupied only seasonally or were built for defense. Whatever the case, they add an intriguing and aesthetically appealing dimension to Sinagua domestic architecture and demonstrate again that there were many ways to be Sinagua.

OBJECTS AND TOOLS

In most cases, only durable materials (such as pottery and stone tools) survive the destructive natural forces acting upon archaeological sites. Occasionally, however, special circumstances such as extreme dryness preserve fragile artifacts, food remains, and other objects from decay or consumption by insects, bacteria, and other destructive agents.

Discoveries at Wupatki Pueblo provide an important and instructive case of special preservation. This site, partially excavated in the 1930s by the Museum of Northern Arizona, has yielded an extraordinary collection of wooden artifacts, textiles, cordage, basketry, and other perishable artifacts. All of these items were preserved in the deep, dry deposits of the pueblo, which had never been penetrated by the sparse rain that falls in the Wupatki Basin.

Among the items recovered were wooden digging sticks, ladles, spoons, pestles, and balls; a bird effigy carved from pine bark; painted prayer sticks; ceremonial canes; arrows; cradleboards; sandals; pieces of blankets fashioned from rabbit fur, turkey feathers, and agave fibers; a belt made from yucca fibers; a yucca fiber bag with a leather

Model of Sinagua dwelling from Walnut Canyon National Monument. Photograph by June Fike

*Anasazi Black-on-white.
Photograph by Gene Balzer*

drawstring; pieces of yucca or beargrass sleeping mats; and numerous pieces of baskets.

The recovery of weaving tools, as well as all parts of the cotton plant, indicates that cotton was cultivated nearby and that it was being spun into fiber and woven into cloth at the pueblo. The range of weaving styles and decorative techniques is impressive and suggests a high degree of sophistication among Wupatki weavers.

In fact, given its low elevation, Wupatki Pueblo may have been one of the few northern Sinagua settlements able to grow cotton successfully. If they did, cotton cloth and perhaps raw cotton may have been an important trade item, accounting for some of Wupatki's apparent prominence as an exchange center.

Other categories of artifacts are preserved at nearly all Sinagua sites. Perhaps the most informative and important of these is pottery. One of the intriguing aspects of the Sinagua is that they apparently chose not to make decorated pots. Except for a few examples of plain brown or red-slipped pots painted with white designs, and an even smaller number of "red-on-buff" vessels that appear to be imitations of Hohokam pottery, the Sinagua had no painted types.

There was, however, no shortage of painted pottery in Sinagua households. By far the majority of this was imported from the Kayenta Anasazi, who lived across the Little Colorado River to the northeast. Decorated Kayenta pottery is present, often in significant quantities, throughout most of Sinagua time. Interestingly, Sinagua pottery is almost completely absent from Kayenta sites, so the exchange obviously did not involve "pots for pots."

Why the Sinagua never developed a significant decorated pottery tradition—and precisely what they might have offered the Kayenta in return for their pots—remain a mystery. Perhaps Anasazi Black-on-white, Black-on-red, and Polychrome pottery, difficult or impossible to make using the raw materials available in much of the Sinagua country, had a symbolic importance for which local products could not be substituted. Or perhaps the exchange of decorated pottery served important social and economic functions, linking the Sinagua with their neighbors and providing a commodity with an agreed-upon value that crosscut different cultural groups. Whatever the reason, there was never a serious attempt to replace Kayenta pottery with any locally made types.

Above: A doorway into the past—a cliff dwelling at Walnut Canyon National Monument. Photograph by Ralph Lee Hopkins

Right: This Kayenta Polychrome pot is a good example of the Anasazi's trade ware. Photograph from MNA Collections by Gene Balzer

HISTORY OF THE SINAGUA

THE cultural traditions of the northern and southern Sinagua apparently developed in parallel but separate ways. Although they shared broadly similar phases of cultural change, each group retained its distinct identity. Thus, it is difficult to detail a general history. Our knowledge is greatest for the Sinagua "heartland," the northern Sinagua territory lying east and south of the San Francisco Peaks, and the following discussion concentrates on that area.

The Earliest Sinagua: Cinder Park Phase
(A.D. 700 TO A.D. 825)

The first Sinagua settlements appeared in the Flagstaff region about A.D. 600. This period, designated the Cinder Park phase, was characterized by small groups of pithouses scattered in a few locations such as the east side of the San Francisco Peaks and the Anderson Mesa area. The Sinagua apparently settled here because of the arable land. Many Cinder Park phase sites are located near alluvial parks (open, grassy areas) in the pine forest. These locations would have provided optimum conditions for floodwater farming. A few settlements also were established along the banks of the Little Colorado River, presumably to take advantage of the agricultural opportunities offered by the floodplain there.

Dwellings of the Cinder Park phase generally consisted of small earth and wood pithouses containing a small central hearth and one or more storage basins beneath the floor. These pithouses often were dome or tipi-shaped, as indicated by inward-slanting wall poles. Exceptionally large and deep pit structures also have been documented. One excavated example was five feet deep and twenty-eight feet in diameter. Structures of this size may have been used for ceremonies that incorporated many families or even the residents of multiple pithouse groups.

Design motifs and colors often crossed cultural boundaries—as these pieces show. Lower left: Elden Corrugated Red-on-buff. Center: Casa Grande Red-on-buff. Lower right: Sacaton Red-on-buff. Photograph by Gene Balzer

Manos, metates, stone hoes, and charred food remains indicate that corn, beans, and squash were being grown, probably through floodwater farming at the mouths of seasonal washes. However, abundant projectile points and animal bones indicate that hunting also provided a major part of the diet. The presence of turkey bones and, in at least one case, an intact turkey skeleton indicates that these birds were hunted or raised for feathers and perhaps for food as well.

Pottery of the Cinder Park phase was usually dominated by an Alameda Brown Ware type called Rio de Flag Brown. Jars and bowls, including small seed jars and canteens, were common. Most household assemblages also included a few vessels of both Anasazi and Cohonina gray ware. Decorated pottery was primarily early Anasazi types such as Lino Black-on-gray. Other items recovered from Cinder Park phase sites include stone axes, both stone and clay pipes, pigments in a variety of colors, crystals, bone awls, polishing stones, and perforated sherd disks used for spinning cotton, yucca, or agave fibers into thread.

Lino gray ware, an early form of Anasazi pottery. Photograph by Gene Balzer

All of this evidence indicates that by around A.D. 700 some groups of people, representing the first recognizable Sinagua populations, had committed to at least a semi-sedentary existence involving the construction of houses arranged into small clusters and the cultivation of corn, beans, and squash. To at least some extent, this settlement pattern depended on stored foods, cached in small pits beneath the floors of pithouses.

The Cinder Park phase Sinagua traded with a number of surrounding groups, primarily the Anasazi but also the Cohonina. Shell and other artifacts indicate trade contacts with the Hohokam as well, either directly or through intermediaries. The presence of a few very large structures suggests that even at this early date the Sinagua had developed some level of religious and perhaps sociopolitical organization incorporating multiple family or community units.

SUNSET AND RIO DE FLAG PHASES (A.D. 825 TO A.D. 1070)

Following the Cinder Park phase, the Sinagua culture underwent considerable growth, elaboration, and diversification. This time of gradual change is divided into two general periods of development—the Sunset (A.D. 825 to A.D. 1000) and Rio de Flag (A.D. 1000 to A.D. 1070) phases. In the space of these four centuries, the Sinagua grew in numbers and expanded their range beyond the alluvial parks that had been the prime focus of Cinder Park settlement. These new areas of expansion included the flanks of Mount Elden, the Walnut Canyon and Rio de Flag drainages, and Medicine Valley.

As groups of Sinagua moved out into new territory, they adapted their farming techniques. Floodwater farming of alluvial parks continued as an important agricultural strategy, but new techniques also were employed. The proximity of many Rio de Flag phase settlements to major washes implies that farming took place in or near the bottoms of those washes. Checkdams and other water control features enabled farmers to slow or divert the runoff from summer thunderstorms, thus preventing crop loss from flooding.

Another agricultural strategy that first appeared during the Rio de Flag phase was the use of "field houses." These structures, usually a single masonry room, probably were used as seasonal dwellings or shelters for farmers tending fields located some distance from their permanent homes. The farming of distant plots evidently reflected a desire to expand and diversify field locations, perhaps because a steadily increasing population was exerting considerable pressure on the limited farmlands available in parks and wash bottoms.

The diversity in pottery types found at both Sunset and Rio de Flag sites suggests that the Sinagua were trading actively with their neighbors. Alameda Brown Ware was the dominant plain ware, but the residents of some settlements also imported significant quantities of Anasazi and Cohonina gray ware pots. Decorated pottery was mostly Anasazi Black-on-white (Kana-a and Black Mesa) and Black-on-red (Deadmans) types, but some settlements also had access to Cohonina Black-on-gray (Floyd and Deadmans) and Hohokam Red-on-buff (Santa Cruz and Sacaton) pottery. All of these imported pottery types probably were obtained in a "down-the-line" fashion, through direct trade with neighboring groups. Thus, the northeastern quadrant of the Sinagua territory contains the highest proportions of Anasazi gray wares, while the northern and northwestern edges have the highest frequencies of Cohonina gray wares, with proportions of these wares diminishing with distance from their source. The result was a mosaic of Sinagua pottery assemblages, with individual sites showing pronounced differences from one end of Sinagua country to the other.

Architectural differences were even more pronounced, especially with houses. Pithouses varied tremendously in size, depth, form, and contents. Most were constructed of wood and earth, but there were also a few masonry-lined examples. New forms of surface architecture also appeared, including masonry or stone and adobe "granaries," apparently used for the storage of corn or other foods.

Some Sunset and Rio de Flag pit structures, like some built during the Cinder Park phase, were exceptionally large. These constructions, easily distinguished in size from the more common pithouses, are probably best interpreted as some sort of ceremonial facility.

Another category of public or ceremonial architecture was the Hohokam-style ballcourt. This feature, clearly derived from the Hohokam living in the Gila and Salt River valleys, may have reached the north via Hohokam trade networks.

The abundance of Kana-a (Anasazi Black-on-white) pottery at both Sunset and Rio de Flag sites suggests that the Anasazi were active trading partners throughout these periods. Photograph from MNA Collections by Gene Balzer

Amphitheater at Wupatki National Monument. Photograph by Dick Dietrich

The Hohokam's Santa Cruz Red-on-buff pottery exemplified by this bowl from MNA Collections was also a popular trading item. Photograph by Gene Balzer

Deadman's Black-on-red pot. Photograph by Gene Balzer

In fact, ballcourts in Sinagua country might have been accompanied by a Hohokam "trader-in-residence" who served as a full-time trade specialist. Another plausible reason for the presence of the ballcourts is the adoption of Hohokam ideology and ceremonies by some segments of Sinagua society.

Toward the end of the Rio de Flag phase, and just before the eruption of Sunset Crater, yet another special architectural form—the masonry "fort"—appeared. These large, thick-walled structures, often surrounded by pithouses and masonry rooms, were distributed from the northern Sinagua frontier westward into the heartland of the Cohonina culture. As their name implies, they were originally thought to have been defensive fortifications, but this interpretation no longer seems valid. Most of the "forts" occur on hilltops or other prominent points along probable transportation routes. Since they contain a wide variety of pottery types, they might have been important features in regional trade. They might also represent ceremonial or communal architecture not yet understood. Whatever they were, these "forts" did not last very long; all well-dated examples span a brief period from about A.D. 1050 to A.D. 1065.

Post Eruptive Patterns

The end of the Rio de Flag phase is closely associated with an extraordinary geological event, the eruption of Sunset Crater. Evidence from archaeology, dendrochronology, and geology shows that the initial eruption took place in A.D. 1064, followed perhaps by a second major eruption in A.D. 1067. In the immediate vicinity of the crater, the eruption brought complete environmental devastation in the form of thick lava flows, poisonous gases, and volcanic bombs. Beyond this zone of total destruction, the landscape was blanketed with deposits of volcanic ash and cinders, ranging from a several feet to only a few inches deep, depending on distance from the crater.

The effect of these eruptions on the Sinagua has been much debated. In the 1930s, Harold S. Colton of the Museum of Northern Arizona proposed that the cinders and ash ejected by Sunset Crater had greatly improved the agricultural potential of a vast area north and east of the volcano by acting as a mulch. According to Colton, the improved farming conditions fostered a substantial increase in local populations—including immigration by Hohokam, Mogollon, Cohonina, and Anasazi peoples. These groups brought with them

In the period after Sunset Crater's eruption, the northern Sinagua population flourished. The Sinagua also expanded their territory— moving onto the piñon-juniper and grassland areas east and north of Flagstaff. Photograph by Gene Balzer

new cultural traits, beliefs, and practices. The result was a transformation of the post-eruptive Sinagua into a culture quite distinct from that of pre-eruptive times.

In recent years, Colton's scenario has been challenged, most notably by Coconino National Forest Archaeologist Peter J. Pilles, Jr. According to Pilles, analysis of Sinagua sites does not support the idea of substantial immigration. Further, many of the cultural changes within the zone of volcanic deposits also took place in areas unaffected by the eruption of Sunset Crater. Thus, Pilles suggests that post-eruptive Sinagua patterns have their roots in a variety of causes—including local environmental changes and region-wide cultural trends during the eleventh and twelfth centuries.

Angell, Winona, and Padre Phases (A.D. 1070 to A.D. 1150)

Following the eruption of Sunset Crater and into the first few decades of the 1100s, Sinagua population increased. This increase in population was accompanied by a gradual movement into lower elevations. Thus, in addition to occupying the parks, mountain flanks, and ridges of the ponderosa pine forests, the Sinagua also established a number of large pithouse and pueblo villages in the piñon-juniper and grassland zones east and north of Flagstaff.

This period also saw intensified regional interaction, with an especially close connection between the Sinagua and the Hohokam. Around or just after the eruption of Sunset Crater, a number of new ballcourts were constructed. The ballcourt sites often show Hohokam traits—including cremation burials, red-on-buff ceramics, clay figurines, shell jewelry, and Hohokam-style pithouses. Many sites also show a strong connection with the Cohonina by the presence of Cohonina gray ware pottery. Contact with the Kayenta and Winslow Anasazi, Mogollon, and Prescott cultures also increased. The specific reasons for the intense interaction are not known, but it clearly follows the Sunset Crater eruption.

The Winona, Angell, and Padre phases span a period of exciting changes that last from about A.D. 1070 to A.D. 1150. Sites such as Winona Village, Ridge Ruin, Juniper Terrace, and Wupatki Pueblo exemplify the trends of this time. Although each of these settlements was established at a different time and each is distinct in form and contents, all share a number of similarities. First, all were founded after the Sunset Crater eruption in environments beyond and below the pine forest zone that was the focus of earlier occupation. Second, each site represents a substantial grouping of people, surrounded by numerous "satellite" pithouses and pueblos that probably were integrated into a single community. Third, each site has at least one Hohokam-style ballcourt nearby. Fourth, all the sites have one or more extraordinarily large subterranean structures that probably served a special function in the larger community. Finally, each exhibits a wide range of exotic goods—including imported plain ware, corrugated, and decorated ceramics, stone and shell jewelry, turquoise, pigments, and other items presumed to have had a high economic or symbolic value.

The location of these sites (along with agricultural features) suggests that the shift to lower elevations was made possible by changes in farming strategies. Essentially, these new techniques emphasized extensive dry farming by seasonal or daily occupation of farmsteads or field houses. Even though field houses are known from the earlier period (and some Sinagua groups continued to farm upper-elevation parks and wash bottoms), dry farming became exceptionally important toward the end of the eleventh and into the early twelfth centuries.

One element of the dry-farming strategy involved continually shifting field locations. Widespread distribution of improvements such as rock piles, terraces, checkdams, and rock alignments and field houses suggests a high turnover rate for individual plots. Such a strategy might have been required by relatively shallow, infertile soils that were easily eroded and exhausted of nutrients once they were cultivated. In some areas, soil erosion might have been somewhat ameliorated by constructing windbreaks or checkdams. The exhaustion of soil nutrients might have been forestalled by adding household trash or rotating crops. In addition, some of the farming pressure on higher-elevation lands probably was relieved by the colonization of the Sunset Crater ash- and cinder-fall zone, which was settled extensively after A.D. 1100.

Whatever the possible problems with such agricultural strategies, Sinagua farmers during the Winona, Angell, and Padre phases were quite

Shell and turquoise necklaces, like these from MNA's Collections, were popular trade items that were both made locally and imported. Photograph by Gene Balzer

successful, and most regions show a population increase throughout these phases. Developments during this period set the stage for the succeeding Elden phase, a time that might well be considered the "Classic" period of Sinagua culture.

THE ELDEN PHASE
(A.D. 1300 TO A.D. 1220)

Sometime around A.D. 1150, a series of new pottery types, principally Flagstaff and Walnut black-on-whites, marked the beginning of what archaeologists call the Elden phase. In reality, except for changes in pottery, there are few valid reasons to draw a sharp boundary between the Padre and Elden phases. The Elden phase in many ways continued and intensified previous trends, such as increasing population, expansion of dry-farming, and construction of multiroomed pueblo communities. In fact, many of the largest and most significant Sinagua settlements—such as Elden, Turkey Hill, Ridge Ruin, Wupatki, Old Caves and New Caves pueblos, and the cliff dwellings of Walnut Canyon—have their origins in the Padre phase, if not before.

Nonetheless, Sinagua accomplishments during the Elden phase were extraordinary, and this phase marked a peak in their culture. Probably the most visible evidence of this success are the large pueblo communities. Although not exceptionally large by some standards, these pueblos indicate an unprecedented level of population concentration. Wupatki, Elden, and Turkey Hill pueblos were among the largest and most complex in their arrangement, housing significant numbers of people. Often, these large central pueblos were surrounded by many dispersed, smaller settlements, a notable aspect of the Sinagua pattern during the Elden phase.

Most such communities were surrounded by extensive agricultural field systems, and Sinagua farming efforts reached their maximum extent during the late Elden phase. Field houses were spread across the landscape, perhaps because of constant shifts in field locations. Many of the field systems associated with field houses were elaborate, with lit-

Above: Petroglyphs, like these adjoining a kiva at Wupatki's Crack-in-rock site, continue to fascinate the monument's many visitors. Photograph by Grazina Sakalas

Below: Elden Pueblo. Photograph by Kathryn M. Wilde

erally hundreds of rock piles, checkdams, and alignments arranged into complex configurations. These constructions, apparently designed to divert runoff and slow erosion, required considerable effort to construct and maintain. Even more impressive, some areas show attempts to increase crop yields through such labor-intensive practices as mulching hillside terraces with household garbage and constructing ridged fields by piling volcanic cinders into artificial dunes up to a few hundred feet long.

Another hallmark of the Elden phase is a vigorous system of inter-regional trade. Given the quantities of goods that apparently moved through Sinagua country, some sites surely emerged as trade centers. Principal candidates for such centers are Wupatki and Elden pueblos, but other settlements no doubt were involved as well. One important trade item was decorated pottery, particularly the finely crafted black-on-white, black-on-red, and polychrome types manufactured by the Kayenta Anasazi. The widespread distribution of these ceramics throughout central and northern Arizona indicates that many thousands of Anasazi vessels were traded. At a minimum, the Sinagua could have offered cotton cloth, macaw feathers, salt, and shell jewelry to their Anasazi neighbors. The Sinagua thus served as important middlemen or brokers of a wide variety of trade items, facilitating the flow of both utilitarian and exotic "prestige" goods between the southern and northern Southwest.

During the Elden phase, new forms of sociopolitical organization arose to deal with the increased population and complexity of settlement and subsistence systems. Some researchers have suggested that after A.D. 1150 the Sinagua developed a "chiefdom" level society, with one leader having the power to schedule and oversee agricultural operations, demand tribute in the form of food and exotic goods, organize trade relations, and make other important community-level decisions. Others have criticized this interpretation as being unwarranted by current archaeological evidence and inappropriate to native southwestern cultures.

Whatever the exact form of Sinagua society, all major Elden phase communities did contain multiple and diverse ceremonial or public architectural forms. These features include ballcourts, community rooms, kivas, and plazas. Each large

Dr. Harold Colton, shown here to the right with John McGregor at the Diablo/Kinnikinnick site, gave the Sinagua the name by which we now know them. Photograph from MNA Photo Archives

pueblo settlement shows unique forms, arrangements, and combinations of such features. The Wupatki community, for example, had a ballcourt; circular and rectangular subterranean structures considered to be kivas or ceremonial rooms; a very large, circular masonry structure that has been called an "amphitheater;" and walled "plazas" adjacent to pueblo room blocks. Elden Pueblo, on the other hand, did not have a ballcourt or any structure comparable to the Wupatki amphitheater, but it did have a very large, centrally located community room.

Although many of these facilities seem best suited for public ceremonies, at least some probably served as private meeting places where important community decisions were made. Who might have made such decisions, whether they were made by consensus, and whether leadership roles were hereditary or achieved are unknown. The increasing size and complexity of Sinagua settlements during the Elden phase, however, implies the need for new decision-making organizations. Kivas, community rooms, and other special structures would have been the likely places where such decisions were made.

We have some hints that the Elden phase occasionally was marred by violent conflict. Some of this evidence is direct, in the form of mass burials and mutilations of human bodies, as observed at Wupatki Pueblo and sites in the Big Hawk Valley. Other clues are less clearcut and rely on functional interpretations of architecture. For example, scattered throughout the Sinagua country are pueblos that appear to have been built with defensive considerations in mind.

Examples include the so-called "fort" sites of Walnut Canyon, New Caves Pueblo, the Citadel, and others built on mesa tops, bedrock escarpments, or other defensible positions. These sites should not be confused with the earlier "forts" of late Rio de Flag times. Many of these sites are accompanied by long, seemingly defensive walls, and the pueblos themselves often show small openings traditionally interpreted as observation holes or openings for shooting arrows.

The causes of such conflicts, if they did indeed occur, remain to be discovered. Given the limited potential of many areas to sustain cultivation, it is possible that competition between communities over farmland was a factor. The conflicts also may have been related to broader-scale phenomena, such as a possible mid-twelfth century Kayenta Anasazi intrusion into the northern Sinagua frontier.

Turkey Hill Phase
(A.D. 1220 to A.D. 1300)

Following the Elden phase, the course of Sinagua history becomes less clear. Sometime after about A.D. 1220, many settlements apparently shrank greatly in size, and some were completely abandoned. Overall, the region's population seems to have decreased, and we see a retraction in the area covered by Sinagua settlements. This poorly known period is called the Turkey Hill phase, named for Turkey Hill Pueblo, one of the few large pueblos that continued to increase in size during the first half of the thirteenth century.

The reason for the apparent drop in Sinagua population is unknown, but by the end of the Elden phase environmental conditions may have deteriorated significantly. A drier climate after A.D. 1220 is one sign, but the high population levels and extensive farming systems of the Elden phase also might have taken a considerable environmental toll. Food and fuelwood requirements of the large Sinagua communities must have been considerable, and there is some evidence that by the early 1200s the farming potential of some areas may have been seriously depleted or exhausted.

With the exception of lower population levels and the abandonment of certain communities, many aspects of the culture during the Turkey

Crack-in-rock, Wupatki National Monument. Photograph by Tom Danielsen

Above: The Citadel, one of the forts that many archaeologists think was constructed during the Elden phase of Sinagua prehistory. Photograph by Gene Balzer

Left: Petroglyph near Palatki ruin. Photograph by Les Manevitz

Hill phase remained similar to the Elden phase. For example, except for the introduction of a few new pottery types, Sinagua material culture, housing, and ceremonial facilities remained substantially the same. However, the Hohokam-style ballcourts were no longer in use. Overall patterns of settlement and agriculture also underwent considerable change. There are many fewer settlements than existed in Elden times, and they are distributed over a much more restricted geographical area. Much of the Wupatki area, for example, was abandoned during the Turkey Hill phase. The widespread use of field houses also came to an end. During this phase, Sinagua farmers apparently abandoned the agricultural strategies that had been so successful during much of the twelfth century and reverted to earlier strategies such as farming alluvial parks and floodplains.

CLEAR CREEK PHASE
(A.D. 1300 TO A.D. 1450)

By about A.D. 1300, the abandonment of most remaining northern Sinagua settlements was complete. In the immediate vicinity of Flagstaff, the only evidence of continuing occupation is at Old Caves Pueblo, an unusual hilltop pueblo built over a set of *cavates* (hollowed out bedrock rooms). Reasons for this abandonment are unknown; suggested causes include human-induced environmental degradation, precipitation shortfalls, and regional shifts in trade networks that left the Flagstaff area relatively isolated.

Above the Mogollon Rim, the Sinagua center of population shifted definitely to the south and east, with large pueblo towns on Anderson Mesa, such as the Pollock site, Kinnikinnick, and Chavez Pass pueblos, becoming the focus of Sinagua culture. These large pueblos, built near reliable springs, seeps, or other permanent water sources, represent a type and scale of settlement previously unknown to the northern Sinagua. The multiple pueblos at Chavez Pass were by far the largest of the Anderson Mesa sites; in fact, they comprised one of the largest late prehistoric settlements in the northern Southwest. There are perhaps more than a thousand rooms within the three major room blocks at the Chavez pueblos. Surrounding mesa slopes are covered with hundreds of agricultural terraces. Polychrome, yellow, and orange ware pottery are relatively abundant—as are shell jewelry, turquoise, and other exotic trade goods. All of these phenomena indicate a complex and closely integrated set of late prehistoric communities, many of which are clearly identified in Hopi legends as the ancestral homes of specific clans.

The period of the large pueblo towns was, however, short-lived. Sometime after about A.D. 1450, all the pueblos within the former range of the northern Sinagua culture were abandoned. Most evidence indicates that the remaining populations moved northward to the Hopi mesas. This migration marked the final chapter in Sinagua prehistory and one of the important events leading to the development of modern Hopi culture.

THE SOUTHERN SINAGUA PATTERN

For a variety of reasons, a description of southern Sinagua history is difficult. We know very little about the early prehistory of the middle and upper Verde River valley, for example, and a number of unresolved issues revolve around the questions of cultural boundaries, migrations, and trade relationships for later periods. Archaeologists have advanced vastly divergent scenarios of this prehistory, each of which postulates varying degrees of influence or influx by neighboring groups.

Generally, the early history of the southern Sinagua seems to have been influenced heavily by the area's Hohokam. Before A.D. 1000, sites throughout the Verde Valley show numerous

Petroglyph at Middle Mesa, Crack-in-rock.
Photograph by Gene Balzer

Southern Sinagua sites built or expanded after A.D. 1300, like Montezuma Castle (above) and Honanki (right), were usually constructed as cliff dwellings in protected places. Photograph above by Les Manevitz. Photograph to the right by Scott S. Warren

Above: Outflow canal from Montezuma Well National Monument.
Right: Cliff House ruins at Montezuma Well National Monument. Permanent sources of water such as the well were important and scarce in Sinagua country.
Photographs by Richard Weston

Hohokam traits—such as Hohokam-style pithouses, red-on-gray or red-on-buff pottery, clay figurines, shell jewelry, and cremation burials. As in the nearby Hohokam heartland of the Gila and Salt River valleys, irrigation canals were an important part of the early southern Sinagua pattern.

After A.D. 1000, cultural diversity in the Verde Valley populations increased. In the upper Verde Valley, Hohokam influence lessened greatly, as red-on-buff ceramics and other traits were replaced by those with a more northerly focus. These changes also were accompanied by a shift in the settlement pattern, with large villages replaced by smaller, more dispersed habitations. Hohokam traits remained dominant in the middle Verde, however, and the size of some pithouse villages seems to have increased.

Chavez Pass ruin. Photograph by Sherry Mangum

Sometime around A.D. 1125, there were pervasive changes in southern Sinagua settlement patterns, architecture, and ceramics. Settlements expanded greatly in geographic range. Lowland areas continued to be occupied, but settlements expanded into upland zones, such as foothills, cliffs, and mesa tops. Architectural forms proliferated—pithouses, small room blocks, and a great number of small cliff dwellings and storage rooms were especially prevalent. There was also a sharp decrease in the number of Hohokam traits. The ceramic complex by this time resembled that of the northern Sinagua, dominated by Alameda Brown Ware and Kayenta Anasazi pottery. The reasons for all of these changes remain obscure. Early archaeologists proposed that Hohokam in the Verde Valley were pushed out by northern Sinagua migrants, who brought such traits as masonry pueblos and Anasazi ceramic types. However, such a migration does not seem to be supported by current evidence. Other explanations involving more gradual change enacted by local populations now seem more likely.

After A.D. 1300, the southern Sinagua joined into a series of large sites, often placed on hilltops or other prominent points. Several large cliff dwellings also were built or expanded during this period—including Montezuma Castle, Honanki, and Palatki. Irrigation canal networks were an important part of the subsistence base, as were elaborate systems of dry and floodwater-farmed fields. Most of these sites show extensive inventories of imported ceramics, many obtained from the Kayenta Anasazi and early Hopi groups living above the Mogollon Rim. Altogether, late southern Sinagua sites are an impressive collection of large pueblo communities that appear well adapted to the rich agricultural environment of the Verde Valley and closely connected to surrounding cultural groups.

The southern Sinagua abandoned these settlements, however, at about the same time that the northern Sinagua were vacating the large pueblo towns on Anderson Mesa. By no later than about A.D. 1450, the Verde Valley was completely abandoned, and the southern Sinagua cultural pattern existed no more. Where the people might have gone also remains a mystery, but it seems reasonable to propose that they too moved northeast to join the northern Sinagua and other groups coming together at the Hopi mesas in the mid- to late fifteenth century.

Wooden sticks often were used in gathering wild foods such as prickley pear. Drawing by Denny Carley

Lomaki ruin, Wupatki National Monument.
Photograph by Gene Balzer

CONCLUSION

For more than 700 years, the Sinagua not only endured but prospered in an environment that was often marginal for agriculture and that occasionally must have seemed excessively cruel and capricious to the humble corn farmer.

This feat in itself makes the Sinagua worthy of our admiration, but we should also remember that their accomplishments far transcended the development of a simple agricultural society. For centuries the Sinagua were literally traders to the Southwest, serving as a conduit between the southern deserts and the Colorado Plateau. Perhaps because of this talent for intercultural transactions, they also managed to achieve a degree of regional influence and cultural complexity matched by few of their neighbors. Therefore, the Sinagua were an essential strand in the complex web of prehistoric life in the Southwest, not only reflecting surrounding cultures but also shaping their form and evolution in ways that we do not fully understand.

By the early 1300s, the Sinagua had abandoned much of the extensive territory they had once inhabited and retreated to a few very large pueblo towns in the Verde River Valley and on Anderson Mesa. Sometime between A.D. 1400 and A.D. 1450, these settlements also were abandoned, and the Sinagua were lost as a distinct entity.

Their material legacy can be found in the abundant ruins that are spread across the landscapes of northern and central Arizona. Their cultural and spiritual legacy is carried forward by the modern Hopi. Both bear eloquent witness to the vitality and to the resilience of the Sinagua as well as to the enduring importance of individual lives of struggle and hope that were lived so long ago.

Box Canyon ruins at Wupatki with the San Francisco Peaks in the background. Photograph by Richard Weston

About the Author

Dr. Christian E. Downum is an adjunct assistant professor in the Department of Anthropology at the University of Arizona. He holds a B.A. in sociology and anthropology from Southwestern College (1979) and M.A. (1981) and Ph.D. (1988) degrees in anthropology from the University of Arizona. Field research spans the Hohokam, Mogollon, and Sinagua culture areas in the United States as well as the Trincheras region of northern Mexico.

Recommended Reading

Agenbroad, Larry D.
    1990 Before the Anasazi. *Plateau* 61(2). Museum of Northern Arizona, Flagstaff.

Anderson, Bruce A.
    1990 *The Wupatki Archeological Inventory Survey Project: Final Report*. U.S. National Park Service, Southwest Cultural Resources Center, Professional Paper No. 35, Santa Fe.

Downum, Christian E.
    1988 *One Grand History: A Critical Review of Flagstaff Archaeology, 1851-1988*. Ph.D. dissertation, Department of Anthropology, University of Arizona, Tucson. Manuscript on file at Arizona State Museum Library, Tucson and Museum of Northern Arizona Library, Flagstaff.

Pilles, Peter J., Jr.
    1979 "Sunset Crater and the Sinagua: A New Interpretation." In: *Volcanic Activity and Human Ecology*, edited by P.D. Sheets and D.K. Grayson, pp. 459-485. Academic Press, New York.

    1981 "The Southern Sinagua." In: People of the Verde Valley, edited by Stephen Trimble, pp. 6-17. *Plateau* 53(1). Museum of Northern Arizona, Flagstaff.

Thybony, Scott
    1987 *Fire and Stone: A Road Guide to Wupatki and Sunset Crater National Monuments*. Southwest Parks and Monuments Association, Tucson.

Managing Editor: Diana Clark Lubick
Editorial Assistant: D.A. Boyd
Editorial Intern: Nancy Cannon
Printing by Land O'Sun
Design by Annette Bird-Bentley
Production by Libby Jennings
Color Separations by Color Masters
Typography by MacTypeNet